Ecobrain of Xolo and friends

story to teach kids about Artificial intelligence, IOT And machine learning

CHAPTER 1 – THE SCIENCE FAIR

"Remember, kids, only four weeks to go until the Science Fair." Miss Smith, the Computer Science teacher, had to raise her voice to be heard over the clamor of the bell and the squeak of thirty pairs of shoes. All the kids in the eighth-grade class raced to get out the door, eager to get to recess. All except for one.

Xolo hung back, taking his time to pack away his books.

Miss Smith noticed him and said, "Do you know what you're doing for the science fair yet, Xolo?"

"I can't decide," said Xolo. "There's just so much I'm interested in. I want to learn more about Artificial Intelligence and Machine Learning. But I'm also really fascinated by the Internet of Things."

"Why don't you make something that combines elements of all three?" suggested Miss Smith.

"Like what?" asked Xolo.

"I don't know. You need to think of a problem. A real problem you've noticed in your day to day life. And then, you need to think: how can I use Computer Science to solve this problem?"

Xolo thought for a moment. Suddenly, he had a brilliant idea.

"Thanks, Miss!" he yelled, throwing his things into his bag and skidding out the door.

"You're welcome, Xolo," laughed Miss Smith, shaking her head.

Xolo raced home as fast as his feet could carry him. He wanted to get to work straight away on his brilliant idea. But first, he'd need some help. He couldn't do this alone. So he stopped off at JT and Zoe's house. JT and Zoe were twins, and they were also Xolo's best friends.

Xolo rang the bell impatiently. JT answered the door.

"What's up?" said JT.

"I've had a brilliant idea for the science fair," said Xolo, catching his breath.

"What's the rush," said JT, "it's still weeks away."

"It'll take us weeks to make my idea," explained Xolo, "so there's no time to lose."

JT called Zoe, and together they ran with Xolo all the way back to his house.

Xolo's Dad had given him a space in the garage, which he called his workshop. It was full of electronics equipment, circuit boards, wires, and more. That is where the three friends sat down to get to work.

"So, don't keep us hanging," said Zoe, "tell us all this amazing idea."

"Well," said Xolo, "Miss Smith said to think of a problem. In our daily lives. And you know the biggest problem we have is Climate Change."

"You're going to solve Climate Change?" Zoe said wide-eyed. "That's your idea for the science fair?"

"No," said Xolo, "well, kinda. You know how there's loads of stuff we can do on a day to day basis to reduce our impact on the environment."

"Yeah, like drive less, buy less meat, turn off the lights," said JT.

"Exactly," said Xolo. "Only it's hard to keep track of all those things at once. So, we are going to invent a device that does all that for us! And we'll call it, 'The EcoBrain!'"

"It's going to be an Internet of Things device that connects to the Internet and talks to all the other smart devices in your house. It's going to use Artificial Intelligence and Machine Learning to decide by itself when to turn off a light, or which is the best mode of transport to take. It will make sure you always make the best decision for the planet!"

"OK," said Zoe.

"I don't get it," said JT.

"Which bit, don't you get?" asked Xolo.

"All of it?" said JT.

"Xolo, you're really smart when it comes to Computer Science. You're always acing Miss Smith class. We want to help you, but you're going to have to help us."

"OK," said Xolo, "let's take it step by step. And I'll teach you guys along the way."

"Great," said Zoe.

"We'll start by building the basic control panel. That will be the brains of the device."

"Good idea," said JT, "start with the brain."

"Then," continued Xolo, "we'll program its Artificial Intelligence, and we'll set it up to get smarter using Machine Learning. Finally, we'll connect it to all the other devices in the house using the Internet of Things."

"Let's get started!" said Zoe.

The three friends looked around Xolo's workshop, which was full of parts from old computers. They found the parts they needed. Using Xolo's soldering iron, they put together a prototype for the EcoBrain's control panel.

"Can anyone find a motion sensor?" asked Xolo.

"I've looked everywhere, but there isn't one here," said Zoe.

"It's OK," said Xolo, "we can print one with my 3D printer."

"There!" said Xolo, "The EcoBrain is all ready to test!"

They pressed the power button. Nothing happened. Xolo typed a few lines of code into his laptop.

"We forgot to close one of the circuits," said Xolo. They quickly soldered a loose connection and ran the test again.

"It powers on!" said Zoe. "Now what?"

CHAPTER 2 – ARTIFICIAL INTELLIGENCE

"Now we need to create the Artificial Intelligence," explained Xolo.

"But, first, you need to teach us all about Artificial Intelligence," said JT.

"OK, guys, I'll teach you everything that I know about AI. That's another way of saying Artificial Intelligence.

You and I are smart, and I don't just mean we're good at Math. What I mean is, as humans, we are *intelligent*. We are able to learn and to apply that learning to new situations so we can solve problems we've never encountered before.

Let me give you an example. Let's imagine you go on holiday to a new city. You need to find your way from the train station to your hotel. But you've never been to this city before, so you don't know the way. But, because you are intelligent, you can use what you've learned already about how to navigate in other cities to find and apply the information you need. You locate a map of the new city and use it to find your hotel.

Computers can also learn and solve problems, but in a different way from humans. Computer intelligence is called Artificial Intelligence, or AI for short. Computers can only use the relationship between facts: that's called logic. Computer scientists use logic to write computer programs to learn by taking in information and then making decisions based on that learning.

But we humans have other skills we use when solving problems. Let's go back to the example of visiting a new city. We might use our imagination to visualize the route options and choose the most interesting one for sightseeing. Or we might use our emotions - we might be feeling nervous about traveling alone, so we decide to take the metro instead of walking. We don't know if computers can be programmed with other human-like abilities like imagination and emotions. Some scientists have started trying to teach computers about feelings and emotions."

"Talking about visiting new cities, I got lost once in New York," said JT.

"For about 30 seconds, then Mom found you feeding your hotdog to a pigeon," laughed Zoe. "So, is AI really new?"

Xolo continued: "No, computer scientists have been trying to make computers with human-like intelligence for more than 50 years. Studies into Artificial Intelligence began in 1950 when a British mathematician called Alan Turning described a test to decide if a computer is intelligent, which is now known as the Turning test. So far, no computer has passed it."

"So does Artificial Intelligence not work, then?" asked Zoe.

"On the contrary! Although it has not reached human-like levels yet, Artificial Intelligence is essential to much of the technology we use today, from phones and video games to online shopping.

Artificial Intelligence is already being used for many different purposes. For example, when you play a video game that uses Artificial Intelligence, the video game studies how you play the game to make it more challenging for you. And there are already driverless cars that use Artificial Intelligence."

"Wow, that's cool!" Said JT, "I wish I had a driverless car!"

"You don't need one; you already have Mom driving you wherever you want to go," joked Zoe. "So how will Artificial Intelligence help our science fair project?"

"I'm glad you asked," said Xolo. "For our EcoBrain to work, it needs to gather information about how we use energy, fuel, and resources around the house, and then learn from this information to help us reduce our impact on the planet. Let's take an example:

We use energy when we turn on the light in a room. So, if we plug the EcoBrain into our lighting system, it can learn about how often and when we turn the lights on and off.

If we add a sensor to the room and connect the sensor to the EcoBrain, then it can sense when there is someone in the room or not. So, it could help us save energy by learning to turn the light out as soon as someone leaves the room."

"Dad is always telling us to turn the light off when we leave the room," said Zoe.

"And you are always forgetting!" said JT.

"So how do we give our EcoBrain some of this fantastic Artificial Intelligence, then, Xolo?"

"Simple," said Xolo, " I just need to program it with some code. I've used a programming language called Python to write my code, which is like a set of instructions for the computer. The code will tell the computer how to learn from the information from the lightbulbs and the room sensor. I've written the code already, so I just need to upload it into the EcoBrain. There, done!"

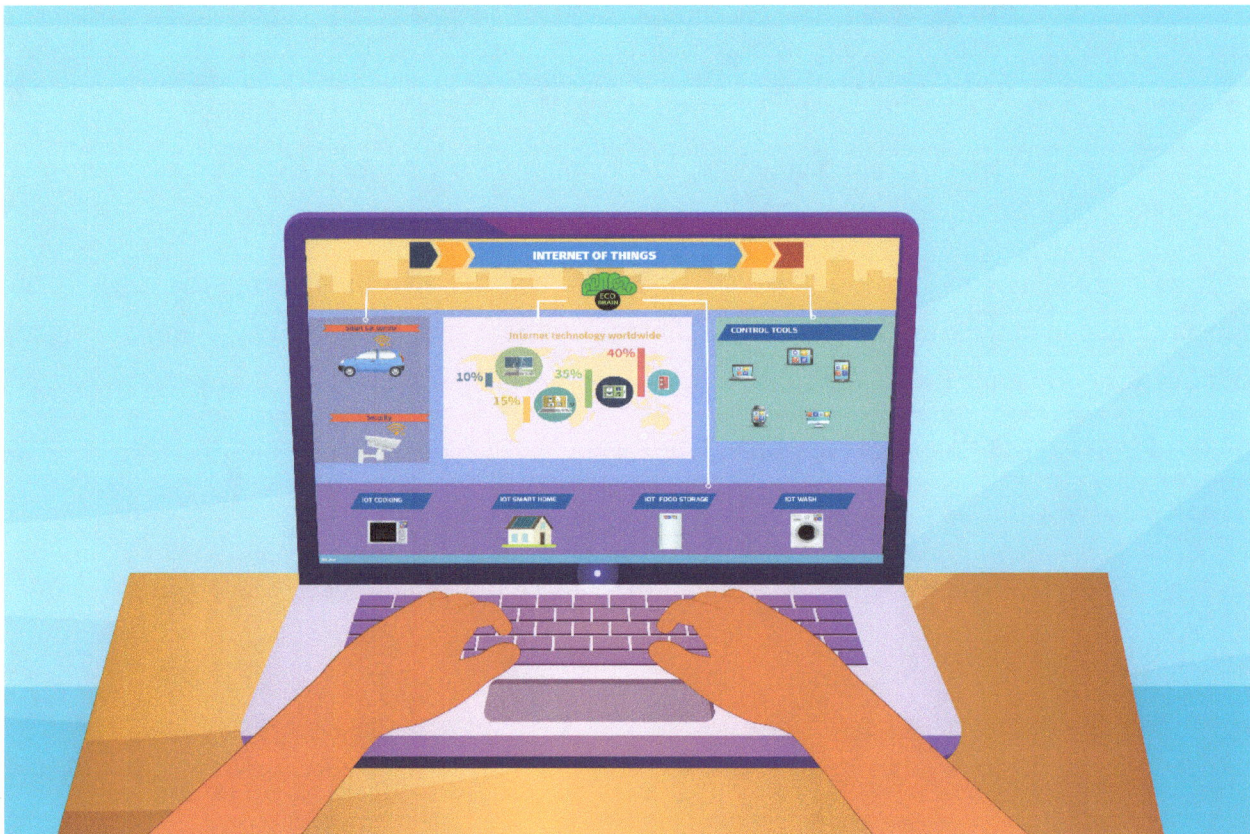

"Great," said Zoe, "what's next?"

"We've built the basics Artificial Intelligence for our EcoBrain. Now, we need it to be able to use Machine Learning," explained Xolo.

"Machine what?" said JT.

CHAPTER 3 – MACHINE LEARNING

"Machine Learning," explained Xolo, "is a special way that a computer can learn to perform a task.

In normal computer programming, if we want a computer to perform a task, we write it a set of instructions to follow, called a program.

Let's use a real example for our science project. What's another big energy drain in your house?" Asked Xolo.

"Heating, I guess," said Zoe, "and Hot Water."

"Great. So, let's look at hot water. We want to have hot water for washing the dishes, for washing hands, for having a bath etc. So, we could just program the EcoBrain to turn the hot water on at the right time for each of these things. But we don't always wash the dishes at the exact time each day. So, instead of using traditional computer programming, we can use Machine Learning.

In Machine Learning, you don't give the computer instructions for how to complete a task. Instead, you feed it lots of examples of that test being completed, and the computer used those examples to learn how to do the task itself."

"That's how I learned to drive," said Zoe.

"What are you talking about? You can't drive!" Said JT.

"I meant, not legally, no. But, I've watched Mom drive so many times that I'm sure I know how to do it!"

Xolo laughed. "Yeah, I guess it is a bit like that. It's a human-like way of learning. So, to train our EcoBrain to save energy on hot water, what do you think we need to do?"

"Train it on lots of examples of us using hot water?" suggested Zoe.

"Exactly," said Xolo. "We need to connect it to the hot water system and then let it learn by observing how we use the hot water over several days or weeks. Then it will learn when we are most likely to need hot water, and it can work out the best way to do that whilst saving energy."

"Awesome," said JT. "The EcoBrain will also be saving money too! Zoe, we should put one in our house and then tell Mom and Dad to give us all the extra money we'll be saving them."

Zoe ignored her bother and asked Xolo, "What other uses are there for Machine Leaning?"

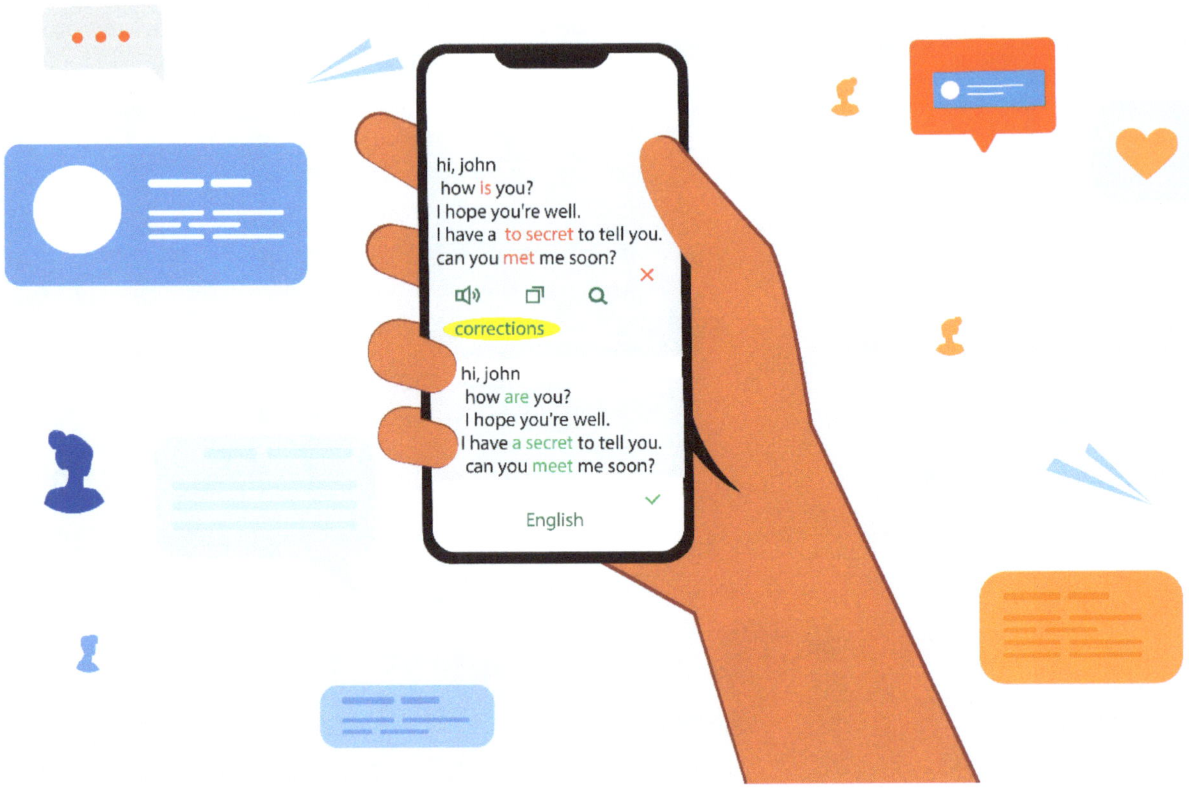

hi, john
how is you?
I hope you're well.
I have a to secret to tell you.
can you met me soon?

corrections

hi, john
how are you?
I hope you're well.
I have a secret to tell you.
can you meet me soon?

English

Xolo explained: "Loads. Think of your phone. When you type a message, the autocorrect fixes any spelling errors you make. It does that using Machine Learning. Do you guys have a digital assistant at home, don't you? That uses Machine Learning too."

"Yeah, Mom keeps it in the kitchen. I accidentally ordered 1000 boxes of chocolate with it the other day. All I said was, I wish I had a lifetime's supply of chocolate. Dad was not impressed."

"I wish we'd brought some with us instead of sending it all back. I'm starving!" said Zoe.

"We need to go to the kitchen in a minute to hook up the EcoBrain to the Internet of Things," said Xolo, "so we can get snacks too. But first, we need to finish programming the Machine Learning element of the EcoBrain. I'll just upload the Python code that I've written so it can learn to control the hot water system. There! Now, let's go connect it to the house and see how it gets on!"

Xolo, Zoe, and JT carefully carried the EcoBrain to the kitchen.

"First things first," said Zoe, reaching for the bread. Working together, they made snacks of sandwiches and fruit.

"Hey, listen to this," laughed JT, "I've written a program for making a sandwich:

 1. Locate the bread packet.

2. Open the bread packet.

3. Remove two slices from the packet ..."

"Funny," said Zoe, "but we could use Machine Learning for this instead. We could feed the computer videos of us making lots of sandwiches, and it could learn from them!"

"Exactly," said Xolo, "you've got it, Zoe."

"A robot that made snacks for us! Now, that would definitely win the science fair prize!"

"That can be our next project," laughed JT.

Their hunger satisfied, the three friends set to work connecting the Eco Brian to the Internet of Things.

CHAPTER 4 – THE INTERNET OF THINGS

"Now I know what the internet is," said JT, but what in the world is the Internet of Things that you keep talking about."

"Put simply, the Internet of Things is about connecting together lots of different kinds of devices using the Internet. Normally, you might just think of the Internet as a connection to your computer or your phone. But you can connect almost anything to the Internet!

Take this fridge, for example. If we make this a smart fridge by putting in a computer chip and a wifi connector, then the fridge can connect to the Internet. Like any computer, it needs its own unique identifier or address. This is called a URL."

"Why would the fridge need to be connected to the internet?" asked JT, "It's not like it needs to chat to its friends!"

"Imagine this. Imagine we put a sensor in the fridge too, so it can sense what is inside it. We connect it via the Internet to our EcoBrain. Now, the EcoBrain can talk to the fridge. I can observe what we put in and out of the fridge, notice how much milk we use each week, for example."

"Then, we could use Machine Learning to teach the EcoBrain how much food we use when!" said Zoe.

"Exactly!" said Xolo. "So then it can make shopping lists for us."

"Or even do the shopping online, like we do at home," said Zoe.

"I get it, I get it," nodded JT. "And it can also make sure we buy exactly what we need, so there is less food waste."

"Precisely," said Xolo.

"How else is the Internet of Things used today?" asked Zoe.

"There are loads of uses, and more are being developed every day. Take that fitness tracker you're wearing, JT, it connects to the Internet, and then you can see information on your computer or phone about how much exercise you've been doing."

"Or, think about a big farm growing wheat for the bread in our sandwiches. The farmer might put sensors around the field to check on weather conditions and how the crops are growing. These could be connected to the Internet. Then the farmer will be able to use the data to manage the farm from his computer."

Zoe looked worried. "But computers can get hacked," he said. "Does it mean that if I connected my fridge to the internet, someone could hack my fridge?"

"Yes, that's one of the issues with the Internet of Things that computer scientists have to work on. Devices need to be as safe as possible from hacking because you don't want someone turning your fridge off in the middle of the night for a prank."

"Let's look around the house for other devices to connect to our EcoBrain," suggested Zoe.

The three friends went on a device hunt, looking for anything and everything they could connect to the Internet and to their EcoBrain project. After a busy hour, they collapsed, exhausted back in Xolo's workshop.

"Right, let's see what we've done," said Xolo. "We've connected the EcoBrain to the hearing system, all the lighting systems, and all the plug points in the house. We've connected it to the fridge, the freezer, the washing machine and the vacuum cleaner. And we've even connected it to my Mom's car. That should be enough to start with."

"Now we just need to check all the programs are running properly, and we are ready to start saving energy!"

Xolo ran some quick systems tests, and thankfully the EcoBrain worked perfectly.

"The science fair is still a few weeks away; we can just sit back and relax now!" said JT.

"I'm afraid not," laughed Xolo, "now, we need to leave the EcoBrain running so it can learn about how we use energy over the next two weeks. Then for the last two weeks, we can let it take charge of all the house systems to see if it really can save energy like we hope it can."

"OK, EcoBrain, let the learning commence!" said Zoe.

CHAPTER 5 – A DREAM COME TRUE

The four weeks until the Science Fair went by in a blur. Xolo and his family carried on life as normal for the first two weeks, going what they normally did around the house. Which included somethings forgetting to turn lights off when they left the room and sometimes buying too much food, so some of it spoiled and went to waste.

After two weeks of data gathering and training, it was time for the EcoBrain to take over. For the next two weeks, the EcoBrain controlled the lights, the hot water, the fridge, and the other devices around the house. It did the grocery shopping online. It even suggested the best routes for Xolo's family to drive, as well as when it would be better to walk or take the bus. Xolo collected data on his family's energy use with and without the EcoBrain. The big questions were: would the EcoBrain save energy? And if so, how much?

Then, the night before the Science Fair, he, Zoe, and JT worked hard to make a poster to communicate their conclusions.

The morning of the Science Fair rolled around, and the three bleary-eyed but excited friends carried their poster and the EcoBrain to school on the bus.

"Welcome, guys!" said Miss Smith, greeting them at the door to the hall. "You're desk is right over there. Go get set up. The judging will start in one hour."

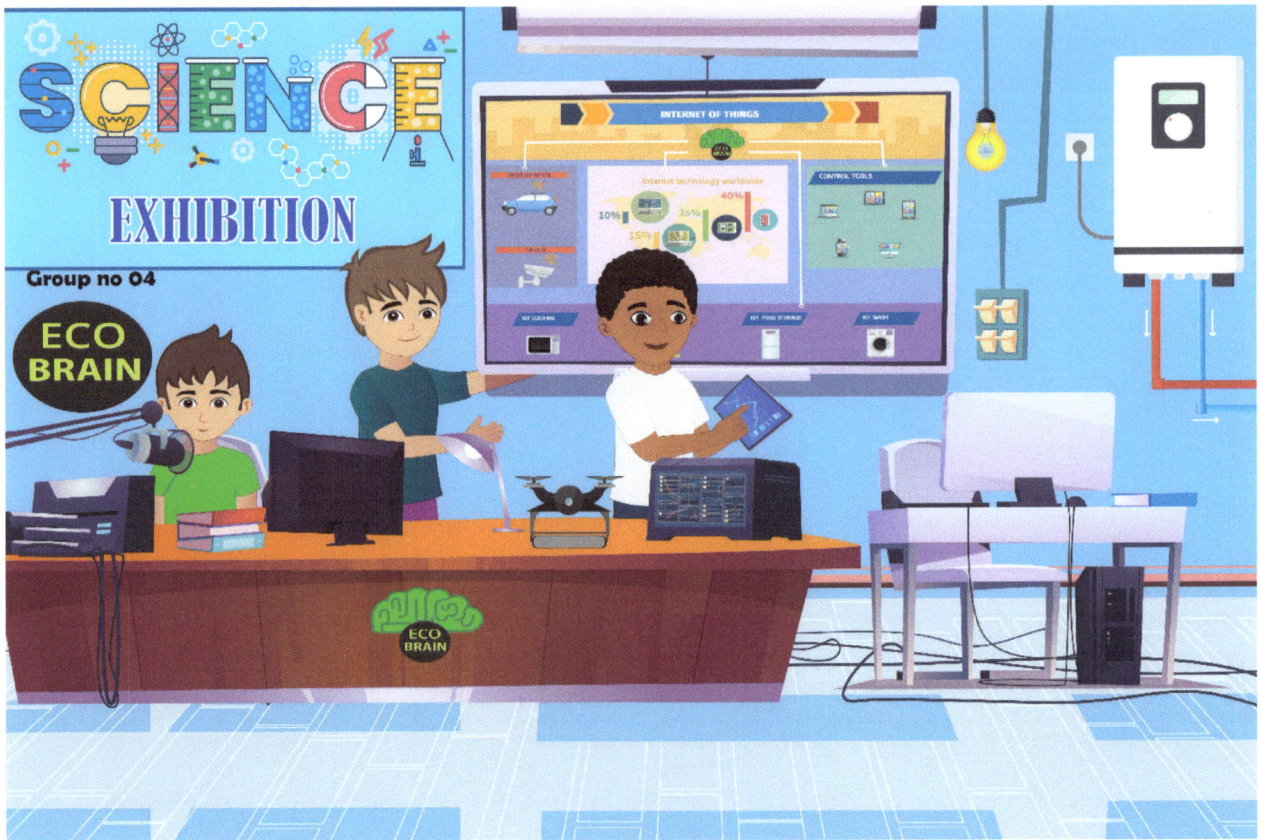

Xolo, Zoe, and JT set up their stand with the EcoBrain in the center, and the posters explain their results all around. Just before the judging was due to start, Miss Smith came around to see them.

"Wow, well done, you guys, this looks awesome," said Miss Smith. "I forgot to say, the judge today is none other than Sia Singh, CEO of EcoCorp."

"What's EcoCorp?" asked Zoe.

"I know EcoCorp," smiled Xolo. They're the biggest developer of environmentally friendly technology in the world. They're in the middle of developing a whole raft of tech solutions to help industry emit fewer greenhouse gases, as well as developing clean energy solutions."

"You know your stuff, Xolo," smiled Miss Smith.

"It's my dream to work for EcoCorp one day," said Xolo.

From across the room, a voice called for Miss Smith. Sorry, kids, I've got to go. Ms Singh is here, and the judging is about to begin."

By now, the hall was full of students and parents. Everyone wanted to have a look at the Eco-Brain and to learn more about Artificial Intelligence, Machine Learning, and the Internet of Things.

In fact, Xolo, Zoe, and JT were so busy that they hardly noticed a woman in a dark business suit stop but their stall with their clipboard. They hardly noticed as the woman intently listened as Xolo explained to a family how he'd written all the Machine Learning code for the EcoBrain. They hardly noticed as the woman saw the Mom and Dad gasped to hear that the EcoBrain had cut Xolo's family energy consumption in half, reducing it by over 50% per week. And they hardly noticed the woman quietly make notes as Xolo described how he envisaged developing and scaling up his EcoBrain.

A bell sounded. Silence fell. Miss Smith's voice carried across the hall. "It is my pleasure to announce the winners of this year's Science Fair."

"Wait, what!" gasped Xolo, "but we haven't even been judged yet!"

"All our hard work," said Zoe, "all for nothing."

"Not for nothing," said JT. "We made an amazing product that cuts home energy use and help to battle climate change."

"You're right, JT, that is reward enough in itself."

Miss Smith continued: "And this year's winner is … the EcoBrain by Xolo, Zoe, and JT!"

The three friends stared at one another, speechless. Slowly they made their way up to the podium, where Miss Smith and Ms Singh stood smiling.

"Congratulations," said Ms Singh, handing over the Gold Science Fair trophy and shaking them by the hand in turn.

"I was incredibly impressed with the EcoBrain's design, especially its use of Artificial Intelligence, Machine Learning, and the Internet of Things," said Ms Singh. "This is exactly the kind of eco-technology that my company, EcoCorp, invents in. As a result, I would like to offer Xolo a job at EcoCorp and change to develop the EcoBrain further in our state of the art labs.

Xolo could not believe his ears. It was his dream come true. To work for EcoCorp! He almost forgot about winning the science fair, until he turned to see the smiling faces of his friends.

"I couldn't have done it without the help of Zoe and JT," he said.

"Of course not. Teamwork is the key," smiled Miss Singh. "That's why we hope you will agree to join the EcoCorp team."

"Yes, please," said Xolo. And the whole hall burst into a round of applause.

ABOUT THE AUTHOR

As a child, Praveen Donepudi dreamed of becoming a great athlete, but instead became well-known in the communications and information tech industries, particularly as an enterprise architect in the IT industry, and he is also a portfolio entrepreneur with interests in multiple small independent businesses. Praveen is also a devoted husband, a loving father, and, of course, a writer.

All of the above keep Praveen's life busy and fulfilling, but that has not stopped him from penning not only the stories that he used to make up to send his daughters off to sleep at bedtime, but also any number of articles, e-books and books, both fiction and non-fiction.

Praveen's work has appeared in such notable international publications as Forbes India, Yahoo Finance, Thrive Global and many more, most often dealing with technology and enterprise architecture based topics, such as the Internet of Things (in which deals with the connectedness of everyday physical objects that form internet-based networks, such as smart homes and similar items), Artificial Intelligence (in which machines work at high rates of cognitive thinking that resemble human logic), machine learning, and neural networks, as well as on finance and marketing. He is also associated with several middleware practices which help applications and operating systems work seamlessly despite differences in age and coding.

Many of Praveen's articles are to be found in Scopus Indexed Journals – a large database of academically solid journals and publications that appear in the form of both ISSN publications (serial publications like journals, magazines and annuals) and IBSN (standalone publications such as books, monographs and the like.) Significant conference papers and seminar presentation papers may also be included – indicating that Praveen's writing is academically rigorous and informatively important.

Not content to sit on his laurels as an author, Praveen gives back too, sitting on the board of a number of journals and providing peer review feedback on submitted pieces.

Away from the office, Praveen loves to travel with his family, seeing all the beauties of nature and the best of mankind and seeing the world not only as it should be, but as it is. He writes for children all over the world in order to impart not only specific knowledge but a love of learning in his readership that will encourage them to ask questions and learn about the world and what they can do to improve on it as well as living out their own dreams.

Praveen's family consists of his beautiful wife, Lakshmi, and his two lively daughters, Shivika and Shanvika, who were his direct inspiration for writing for children. Making up on-the-spot stories for his girls boosted his creativity and Praveen began to jot the stories down with a view to publishing them: yet another goal he has achieved very successfully!

Praveen's stories always hide a moral or message of being positive and enduring the bad times in order to enjoy the good, but they are often so skillfully woven into the tale that the reader is hardly aware that they are receiving an education at the same time as enjoying an entertaining yarn!